# How to be Brilliant at
# USING A CALCULATOR

Beryl Webber
Terry Barnes

We hope you and your class enjoy using this book. Other books in the series include:

If you would like further information on these or other titles published by Brilliant Publications, please write to the address given below.

Published by Brilliant Publications,
The Old School Yard, Leighton Road, Northall, Dunstable,
Bedfordshire LU6 2HA

Written by Beryl Webber and Terry Barnes
Illustrated by Kate Ford
Cover photograph by Martyn Chillmaid

The publishers and authors are grateful to the pupils and staff at the
Edward Clements School in Coventry for their help in trying out the
activities.

Printed in Malta by Interprint Ltd.

# Contents

# Introduction

*How to be Brilliant at Using a Calculator* contains 42 photocopiable sheets for use with 7–11 year olds. The ideas are structured in line with the National Curriculum programmes of study and attainment targets. They can be used whenever the need arises for particular activities to support and supplement whatever core mathematics programme you use. The activities provide learning experiences which can be tailored to meet individual children's needs.

The activities are addressed directly to the children. They are self-contained and many children will be able to work with very little additional support from you. You may have some children, however, who have the necessary mathematical concepts and skills, but require your help in reading the sheets.

The children will need pencils and should be encouraged to use the sheets for all of their working. Obviously, each child will require a calculator for their own personal use. Some activities require additional basic classroom mathematics resources such as dice, counters and interlocking cubes. Some of the sheets will require the use of additional resource sheets and these can be found at the back of the book. Where this is the case, it has been indicated by a small box, with the relevant page number in it, in the top right corner, eg 47 .

*How to be Brilliant at Using a Calculator* relates directly to the programmes of study for attainment target 1 (Using and Applying Mathematics), 2 (Number), 3 (Algebra) and 5 (Data Handling). The page opposite gives further details and on the contents page the activities are coded according to attainment target(s) and level(s). Children working at level 2 will be able to use some of the level 3 sheets but may need additional help from you. Each activity provides experiences to support the acquisition of skills for using a calculator. Many relate directly to attainment target 1 (Using and Applying Mathematics).

Pages 45 and 46 provide self-assessment sheets so that children can keep a record of their own progress.

*Calculators*
As calculators operate in different ways, it is essential that you are familiar with the machine and the activities. The keys depicted in this book are those most frequently found on basic classroom calculators. It is beneficial in the early stages for each child in the group to use the same model of calculator. With a mixture of calculators it is possible to obtain different answers for a single sum. This is because most calculators in primary schools use arithmetic logic and calculate as they go along. More sophisticated calculators use algebraic logic and multiply and divide before they add and subtract. Take the following sum, for example:

$$18 \quad - \quad 2 \quad \times \quad 4 \quad =$$

Arithmetic solution: 64    Algebraic solution: 10

$$18 - 2 = 16$$
$$16 \times 4 = 64$$

$$2 \times 4 = 8$$
$$18 - 8 = 10$$

# Links to the National Curriculum

The activities in this book allow children to have opportunities to:

## Ma2 Number

### Using and applying number
1    Pupils should be taught to:
a    make connections in mathematics and appreciate the need to use numerical skills and knowledge when solving problems in other parts of the mathematics curriculum
b    break down a more complex problem or calculation into simpler steps before attempting a solution; identify the information needed to carry out the tasks
c    select and use appropriate mathematical equipment, including ICT
d    find different ways of approaching a problem in order to overcome any difficulties
e    make mental estimates of the answers to calculations; check results
g    use notation diagrams and symbols correctly within a given problem
h    present and interpret solutions in the context of the problem
j    understand and investigate general statements
k    search for pattern in their results; develop logical thinking and explain their reasoning.

### Numbers and the number system
2    Pupils should be taught to:
b    recognize and describe number patterns, including two- and three-digit multiples of 2, 5 or 10, recognizing their patterns and using these to make predictions; make general statements, using words to describe a functional relationship, and test these; recognize prime numbers to 20 and square numbers up to 10 x 10; find factor pairs and all the prime factors of any two-digit integer.
d    understand unit fractions then fractions that are several parts of one whole
f    recognize the equivalence between the decimal and fraction forms
i    understand and use decimal notation for tenths and hundredths in context.

### Calculations
Pupils should be taught to:
a    develop further their understanding of the four number operations and the relationships between them including inverses; use the related vocabulary; choose suitable number operations to solve a given problem, and recognize similar problems to which they apply
i    use written methods to add and subtract positive integers less than 1000, then add and subtract numbers involving decimals; use approximations and other strategies to check that their answers are reasonable
j    use written methods for short multiplication and division by a single-digit integer of a two-digit then three-digit then four-digit integers; use approximations and other strategies to check that their answers are reasonable
k    use a calculator for calculations involving several digits, including decimals; use a calculator to solve number problems; know how to enter and interpret money calculations and fractions; know how to select the correct key sequence for calculations with more than one operation.

**Solving numerical problems**

4      Pupils should be taught to:

a      choose, use and combine any of the four number operations to solve word problems involving numbers in 'real life', money or measures of length, mass, capacity or time, then perimeter and data.

b      choose and use an appropriate way to calculate and explain their methods and reasoning

c      estimate answers by approximating and checking that their results are reasonable by thinking about the context of the problem, and where necessary checking accuracy

d      recognize, represent and interpret simple number relationships, constructing and using formulae in words then symbols.

## Ma3 Shape, Space and Measures

**Understanding measures**

4      Pupils should be taught to:

a      convert one metric unit to another.

## Ma4 Handling data

**Processing, representing and interpreting data**

2      Pupils should be taught to:

b      construct and interpret frequency tables, including tables for grouped discrete data.

---

On the contents page each activity has been coded to indicate its main relationship with the programme of study for Key Stage 2.

The coding operates as follows:
    Ma2 –     Numbers
    Ma3 –     Shape, Space and Measures
    Ma4 –     Handling data

These codes are followed by a number and lower case letter to indicate the relevant sub-section and aspect.

For example:
Ma2–3(k) indicates Number, sub-section 3 (Calculations), k – 'use a calculator for calculations involving several digits…'.

Each activity is also coded by an upper case letter (A–C) indicating the relative difficulty of the activity itself. Activities coded 'A' are the most challenging.

# Make a number

My number is 451!

You can make any number you like using only these four keys!

| 1 | 0 | + | = |

Give yourself a three-digit number by throwing a die three times: first for the hundreds, then for the tens and finally for the units.

Using these four keys    | 1 | 0 | + | = |    get the calculator to show your number.

How many goes does it take?

> **Tip**: Start with the hundreds number, then add the tens number and so on.

Choose some more three-digit numbers with a friend. Record your numbers here. See who can make the numbers in the least amount of goes.

Now try some four-digit numbers.

---

**EXTRA!**
If you are feeling adventurous, try to make your phone number!

---

How to be Brilliant at Using a Calculator

# Dividing

Dividing is easy with a calculator but you must be careful about the order in which you press the keys. Try this:

this part means
'divided by 9'

Sometimes division sums are written another way.

9 ⟌ 27     which means     27 ÷ 9

If you mix up the numbers you will get a strange answer. Try:

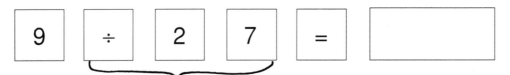

this part means
'divided by 27'

Work these out using your calculator:

125   ÷   5   =                    6 ⟌ 564

10 ⟌ 210                           11 ⟌ 3069

7 ⟌ 3661                           9 ⟌ 8982

216   ÷   4   =

64 ÷ 8 =

**EXTRA!**
Make up a number puzzle using
division sums. Ask a friend
to solve it using a calculator

How to be Brilliant at Using a Calculator

# Cross numbers

Cross numbers are like crosswords, but instead of letters the answers are numbers.
Use your calculator to solve this puzzle.

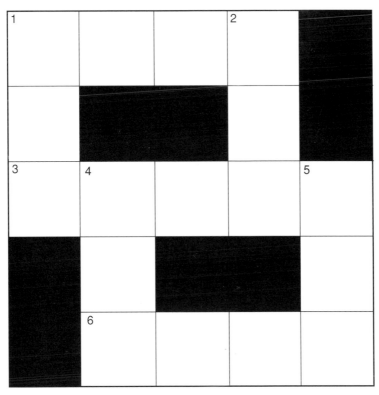

## Across

1   one thousand six hundred and seven
    multiplied by three

3   eight hundred and seventy nine
    multiplied by twenty five

6   twelve multiplied by twelve multiplied
    by ten

## Down

1   twenty four multiplied by eighteen

2   eight hundred and eighty two divided by
    six

4   eleven squared (eleven multiplied by
    eleven)

5   five thousand five hundred divided by
    eleven

---

### EXTRA!
Make up some puzzles of your own. Ask a friend to try them out.

**Tip**: To make up your own cross number puzzle, write numbers in a grid
first. Use your calculator to get to these answers. Write out the clues
and then make another copy of the grid without the answers written in.

---

How to be Brilliant at Using a Calculator

# Secret numbers

Work with a friend. You will need a 100 square (page 47). Secretly choose a number and write it down on the back of this sheet. Don't let your friend see what you have written!

Your friend needs to guess what your number is in as few tries as possible. She guesses a number and enters it into the calculator. You say if it is too high or too low. Your friend chooses another number and uses the calculator to add or subtract it from her first number. You write down each guess. Keep going till your friend gets to your number.

| For example: | Secret number is **67** | |
|---|---|---|
| *Guess 1* | 50 | too low! |
| *Guess 2* | + 14 = 64 | too low! |
| *Guess 3* | + 20 = 84 | too high! |
| *Guess 4* | −10 = 74 | too high! |
| *Guess 5* | − 5 = 69 | too high! |
| *Guess 6* | − 2 = **67** | well done! |

**Tip**: You do not need to clear the calculator between each guess – just carry on.
For example:

| 50 | + | 1 | 4 | = | 64 | + | 2 | 0 | = | 84 |
|---|---|---|---|---|---|---|---|---|---|---|

Have five more goes each. Record the number of guesses here:

| Name: | Name: |
|---|---|
| *Game 1* Number of guesses | *Game 2* Number of guesses |
| *Game 3* Number of guesses | *Game 4* Number of guesses |
| *Game 5* Number of guesses | *Game 6* Number of guesses |
| *Game 7* Number of guesses | *Game 8* Number of guesses |
| *Game 9* Number of guesses | *Game 10* Number of guesses |
| Total number of guesses | Total number of guesses |

# Tens ladder

Cut up a 100 square (page 47) and put the squares in a bag. In turns with a friend, take a number and round it up or down to the nearest 10. Place the number on the correct rung of your ladder.

You need to collect one number for each rung. If you already have a number on the rung you may decide to keep it or swap it for the number you have just picked.

_____'s ladder          _____'s ladder

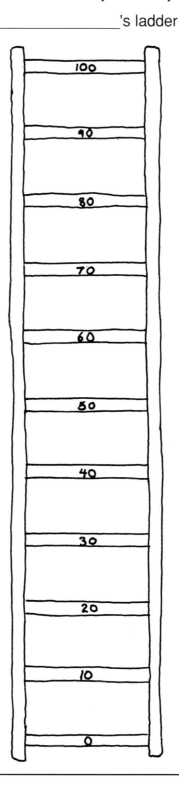

When you have filled all the rungs on your ladder, add up all your numbers using a calculator. The winner is the person who has the highest total.

Play the game several times. What do you notice about the totals?

**Tip**: To decide which 10 to round a number to, look at the units digit:

- if it is 1, 2, 3 or 4 round down to the 10 below.

- if it is 5, 6, 7, 8 or 9 round up to the 10 above.

For example:
- 24 rounds down to 20
- 25 rounds up to 30

**EXTRA!**
Find out the highest and lowest totals possible.

How to be Brilliant at Using a Calculator

# Make my number

Work with a friend. You will need a 100 square (page 47) and a calculator between you.

*It's cheating to use the C and CE keys!*

◆ Player 1 chooses a number.

◆ Player 2 enters it into the calculator and crosses it off the 100 square.

Then Player 2 chooses a different number for Player 1.

◆ Player 1 tries to make the calculator show the new number using the least number of key presses possible.

When he gets it, he crosses the number off the 100 square and chooses a new number for Player 2 to make.

Continue for 10 goes each. Record what you do here. The winner is the player who uses the least number of goes.

---

**EXTRA!**
Try this activitiy without using addition.
It's much harder!

---

How to be Brilliant at Using a Calculator

# Target one

You will need one die numbered 1 – 6.

Enter a number between 10 and 60 into your calculator. Throw the die. You are allowed to use the number you have thrown and the following keys only:

| + | − | X | ÷ | = |

Your target is to get the calculator to display 1, or as near to 1 as you can get.

Record the steps you use (you may not use more than 10 steps).

For example:
Number chosen      56
Number on die      3

| *Calculator display* | | *Keys pressed:* |
|---|---|---|
| Step 1 | 56 | − 3 3 = |
| Step 2 | 23 | x 3 = |
| Step 3 | 69 | + 3 = |
| Step 4 | 72 | ÷ 3 = |
| Step 5 | 24 | + 3 + 3 = |
| Step 6 | 30 | ÷ 3 = |
| Step 7 | 10 | ÷ 3 = |
| Step 8 | 3.3333333 | ÷ 3 = |
| Step 9 | 1.1111111 | |

> **Tip**: You may use more than one operation key in each step, but be careful about the order in which you press the keys.
>
> If you get a negative number you may wish to start again.

---

**EXTRA!**
Try again using a new number. Compare your answer with
a friend's and discuss your solutions.

---

# Clearing

Calculators usually have two methods of clearing the display.
One clears all the numbers you have been using. It may be labelled:

| AC | or | ON/C |   It is called 'All clear'.

The other button just clears the last number you entered. It may be labelled:

| C | or | ON/C | or | CE |   It is called 'Clear error'.

This button is useful if you make a mistake because you don't have to lose all the work you've done already. You can just carry on from where you were before.

Investigate your calculator and find these two buttons.

Try these calculations to see what happens:

| Calculation | Result |
|---|---|
| 4 2 + 6 3 = All clear | |
| 4 2 + 6 3 All clear | |
| 4 2 + 6 All clear | |
| 4 2 + All clear | |
| 4 2 + 6 3 = Clear error | |
| 4 2 + 6 3 Clear error | |
| 4 2 + 6 Clear error + 8 = | |
| 4 2 + Clear error | |

**EXTRA!**
Different calculators may get different results. Compare two different types
of calculator to see if they work the same.

# Calculator logic

Most calculators work out the answer as they go along. Follow the flow diagram and record the display at each stage.

*Display*

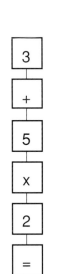

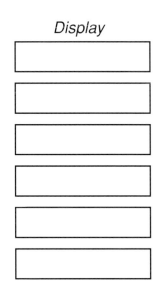

What has the calculator done?

In which order did it add and multiply?

Mathematicians always calculate multiplication and division before addition and subtraction, so they would do this:

3   +   5   x   2   =

3   +    10    =   13

Many calculators don't do this so we have to change the order of the sum, like this:

5   x   2   +   3   =

Work these sums out the way a mathematician would using your calculator. Remember to multipy and divide *before* you add and subtract.

6   +   2   x   9   =        11   –   9   ÷   3   =

21   –   4   x   3.   =        19   +   10   ÷   5   =

31   +   4   x   8   =        27   –   16   ÷   2   =

99   –   50   x   1   =        42   +   15   ÷   3   =

121   +   4   x   0   =        54   –   64   ÷   8   =

How to be Brilliant at Using a Calculator

# Target three

Enter 3 into your calculator. Follow the snake around and return to 3 using the 5s in the snake. You must use the following keys at least once each:

| + | − | X | ÷ | = |

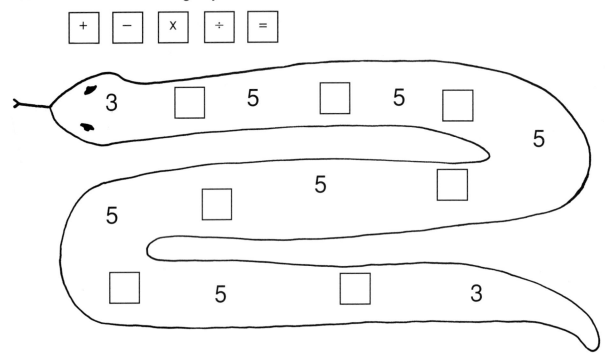

Have another go using this snake:

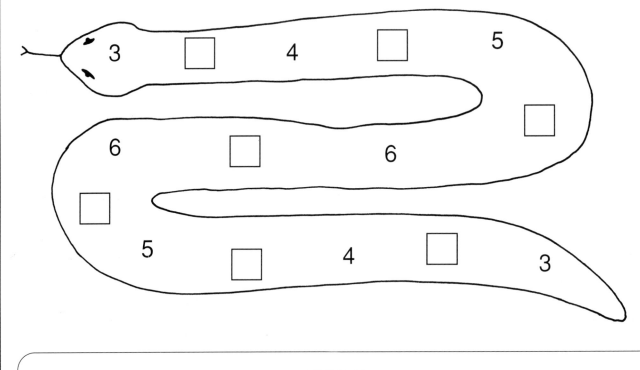

---

**EXTRA!**
Make up some more snakes of your own.

---

# Think of a number

Follow the flow charts. Which ones take you back to your original number?

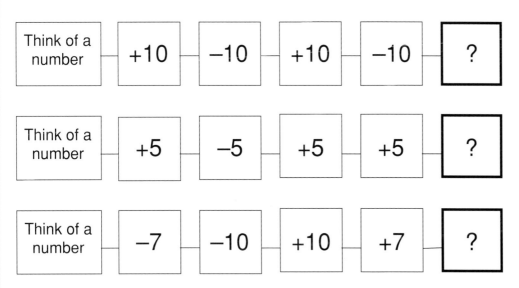

| Think of a number | +10 | −10 | +10 | −10 | ? |
| Think of a number | +5 | −5 | +5 | +5 | ? |
| Think of a number | −7 | −10 | +10 | +7 | ? |

What do you notice about the flow charts that take you back to your first number?
Make up some flow charts of your own that take you back to the number you thought of first.

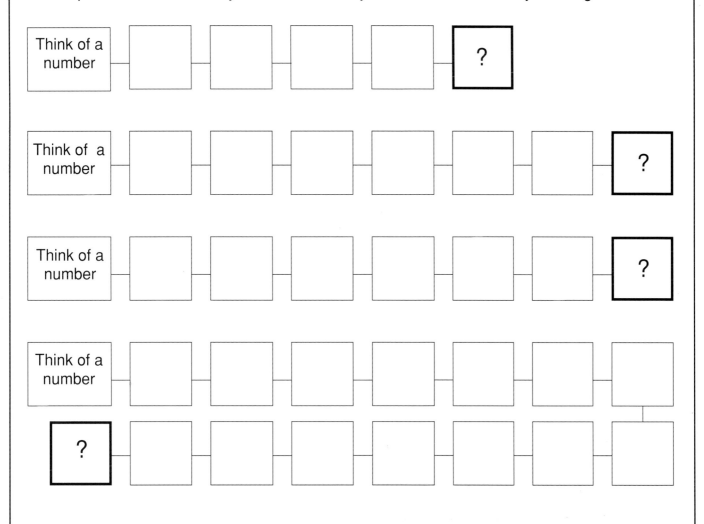

How to be Brilliant at Using a Calculator

# Sequences

You will need some sequence cards (page 48) for this activity.

Look at this flow chart. Use your calculator. Choose a number between 1 and 10 and enter it into your calculator.

Follow the instructions and then press the = key.

Write down the answer in the table below.

Try again with another number.

```
┌─────────────┐
│  Enter a    │
│  number     │
└─────────────┘
       │
┌─────────────┐
│  Multiply   │
│  by 2       │
└─────────────┘
       │
┌─────────────┐
│  Subtract   │
│  2          │
└─────────────┘
       │
┌─────────────┐
│ Record the  │
│ result      │
└─────────────┘
```

Use all the numbers between 1 and 10. Record the results in this table:

| Starting number | 1 | 2 | 3 | 4 | 5 | 6 | 7 | 8 | 9 | 10 |
|---|---|---|---|---|---|---|---|---|---|---|
| Result | 0 | | | | | | | | | 18 |

What do you notice about the results?

Make up your own flow charts using the sequence cards. Try them with numbers 1 – 10 and record the results in a table.

Swap the two middle cards so that the third card is second and the second card is third. Try numbers 1 – 10 again and record the results. What do you notice?

---

**EXTRA!**
Make up some flow charts with three or four middle cards.
Change the cards around. What do you notice?

---

# Volumes

To find the volume of a box you need to meaure the height, width and length.

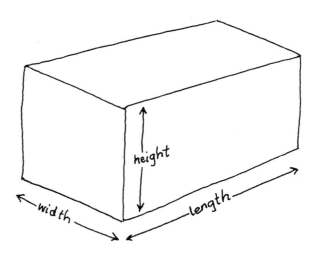

Then you use the formula:

Volume = height x width x length

or

$V \quad = \quad h \quad \times \quad w \quad \times \quad l$

Find the volume of this box.

> **Tip**: The volume is measured in cubic centimetres ($cm^3$).

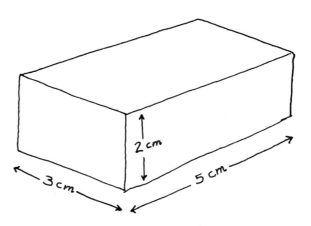

Make this shape with centimetre cubes and count them to check your calculation.

**EXTRA!**
Measure five boxes in the classroom and use the formula to find their volumes.

How to be Brilliant at Using a Calculator

# Fractions

To change from vulgar fractions (eg $\frac{3}{4}$) to decimal fractions (eg 0.75)
you can use the $\div$ key.

For example:

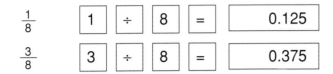

| | | | | |
|---|---|---|---|---|
| $\frac{1}{8}$ | 1 | ÷ | 8 | = 0.125 |
| $\frac{3}{8}$ | 3 | ÷ | 8 | = 0.375 |

Find the decimal fractions for the rest of the eighths family:

| Vulgar fraction | Decimal fraction | Vulgar fraction | Decimal fraction |
|---|---|---|---|
| $\frac{1}{8}$ | | $\frac{5}{8}$ | |
| $\frac{2}{8}$ | | $\frac{6}{8}$ | |
| $\frac{3}{8}$ | | $\frac{7}{8}$ | |
| $\frac{4}{8}$ | | $\frac{8}{8}$ | |

Sometimes you can see interesting patterns. Try the ninths pattern:

| Vulgar fraction | Decimal Fraction | Vulgar fraction | Decimal Fraction |
|---|---|---|---|
| $\frac{1}{9}$ | | $\frac{6}{9}$ | |
| $\frac{2}{9}$ | | $\frac{7}{9}$ | |
| $\frac{3}{9}$ | | $\frac{8}{9}$ | |
| $\frac{4}{9}$ | | $\frac{9}{9}$ | |
| $\frac{5}{9}$ | | $\frac{10}{9}$ | |

**EXTRA!**
Try the sevenths pattern.
What happens when you continue this pattern?

# Square numbers

Use your calculator to work out the squares of numbers from 1 – 20. Record them in the table.

**Tip**: To work out the square of a number either

| enter it | x | = | or |

| enter it | x | x | = |

A square number is what you get when you multiply a number by itself.

| Number | 1 | 2 | 3 | 4 | 5 | 6 | 7 | 8 | 9 | 10 | 11 | 12 | 13 | 14 | 15 | 16 | 17 | 18 | 19 | 20 |
|--------|---|---|---|---|---|---|---|---|---|----|----|----|----|----|----|----|----|----|----|----|
| Square |   |   |   |   |   |   |   |   |   |    |    |    |    |    |    |    |    |    |    |    |

Use your calculator to add up the consecutive odd numbers. Record the totals as you go. What do you notice about the pattern?

| 1 | = | |

| + | 3 | = | |

| + | 5 | = | |

| + | 7 | = | |

| + | 9 | = | |

| + | 11 | = | |

| + | 13 | = | |

| + | 15 | = | |

| + | 17 | = | |

| + | 19 | = | |

| + | 21 | = | |

| + | 23 | = | |

| + | 25 | = | |

| + | 27 | = | |

| + | 29 | = | |

**EXTRA!**
Predict what the total of the first 20 odd numbers will be. Try it and see.

How to be Brilliant at Using a Calculator

# Dot to dot

Press either

7  +  =    or

7  +  +  =    to set up a constant addition function to add 7.

Keep pressing  =    and join each number you find in the number chain to reveal a picture.

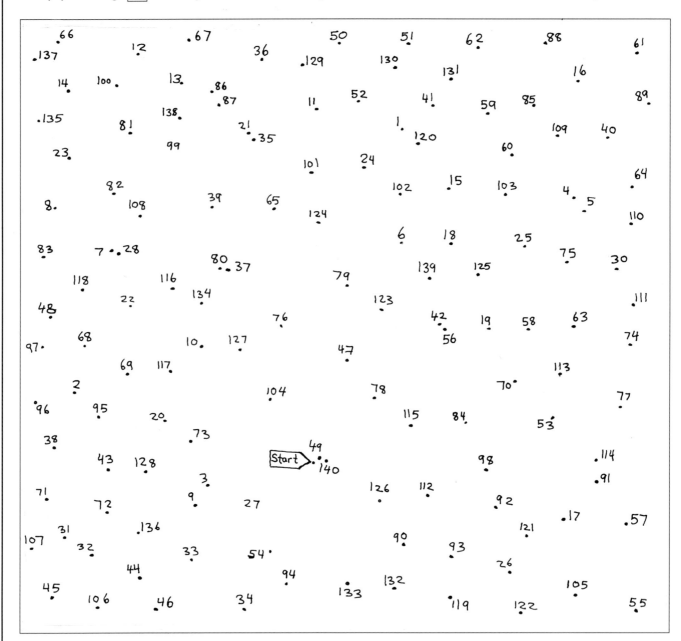

---

---

# Balloons

Colour the balloons showing numbers for the **3** times table red.
Colour the balloons showing numbers for the **4** times table yellow.
Colour the balloons showing numbers for the **5** times table blue.
Colour the balloons showing numbers for the **6** times table green.

If you have already coloured a balloon and its number comes into another table, add some spots in the second colour. For example, 12 would be red with yellow and green dots.

**Tip**: Set up a constant addition function on your calculator to help you decide which balloon is in which table. For example, to find the numbers in the 3 times table, press either:

| 3 | + | = | = | = |  etc.  or

| 3 | + | + | = | = | = |  etc.

Change the start number for the other times tables.

# Changing units

Don't forget to round your numbers to the nearest centimetre or metre.

To change millimetres to centimetres you can set up a constant division function to divide by 10 by pressing either:

| 1 | 0 | ÷ | = |    or

| 1 | 0 | ÷ | ÷ | = |

To change centimetres to metres you can set up a constant division function to divide by 100.

Complete the table below using constant division functions. Round your numbers to the nearest centimetre or metre.

| mm | cm | m |
|---|---|---|
| 175 | | |
| 783 | | |
| 1429 | | |
| 1976 | | |
| 2430 | | |
| 2761 | | |
| 5919 | | |
| 6999 | | |

---

**EXTRA!**
How would you change metres to kilometres?
What constant function could you use?

---

# Investigating number patterns

Use your calculator to investigate these patterns. Predict the next few steps.

1        x    =

11       x    =

111      x    =

1111     x    =

> **Tip**: You will need to set up a constant multiplication function. You may need to use these keys:
>
> | x | | x | | = |

What happens if you try the next step on your calculator?

9        x    =                    3        x    =

99       x    =                    33       x    =

999      x    =                    333      x    =

9999     x    =                    3333     x    =


1        ÷    =                    9        ÷    =

10       ÷    =                    99       ÷    =

100      ÷    =                    999      ÷    =

1000     ÷    =                    9999     ÷    =


## EXTRA!
Make up some similar patterns of your own.

# Sandwich cake

**Sandwich cake**
Serves 6

You will need:
100g butter or margerine
100g caster sugar
2 eggs
100g self-raising flour
1 tablespoon hot water

Beat the butter and sugar together until light and fluffy. Beat in the eggs, one at a time. Fold in the flour, then the hot water.

Pour the mixture into 2 greased and lined 18 cm cake tins. Bake in an oven at 180 C (350 F, gas 4) for 20-25 minutes. Turn onto a wire rack to cool. Sandwich the two halves together with jam.

Recipes usually tell you how many people can be served. Sometimes you may want to make enough for more people. Then the amount of each ingredient must be increased. To make enough sandwich cake for 12 people you would need to double all the quantities. To make enough for 9 people you would need to multiply each quantity by 1.5.

Set up a constant multiplication function to multiply by 1.5 and work out how much of each ingredient you would need.

Write the ingredients for making enough sandwich cake for 9 people here.

**Tip**: To set up a constant multiplication function to multiply by 1.5 press either:

| 1 | . | 5 | x | = |   or

| 1 | . | 5 | x | x | = |

## EXTRA!
Try changing the number of people this orange cream recipe serves. You could make it smaller by setting up a constant division function.

**Orange cream**
Serves 6

1 x 142g packet orange jelly
1 orange
1 x 175g can evaporated milk
225g full fat soft cheese

Make the jelly according to the instructions on the packet. Grate the rind from the orange and save for decoration. Squeeze the juice from the orange. Whisk the made-up jelly, the freshly squeezed orange juice, the evaporated milk and the full fat soft cheese together. Pour the mixture into a serving dish and leave it to set in the refrigerator. Decorate with the orange rind when set.

# Finding prime numbers

You will need a copy of a 100 square (page 47).

Prime numbers are those that can only be divided by 1
and themselves and give a whole number answer.

For example:

◆   3 can be divided by 1 and 3 only to give a whole number. It is a prime.

◆   4 can be divided by 1, 4 and 2 to give a whole number, so it is not prime.

Use your calculator to help you find out whether each of numbers 1 – 10 is prime or not.
Colour the prime numbers in on ten grid below.

| 1 | 2 | 3 | 4 | 5 | 6 | 7 | 8 | 9 | 10 |
|---|---|---|---|---|---|---|---|---|----|
|   |   |   |   |   |   |   |   |   |    |

> **Tip**: If a number is the answer for a
> multiplication sum, it cannot be prime. Set up
> constant functions on your calculator to find the
> multiples.

Record all the prime numbers up to 100 on the 100 square.

**EXTRA!**
Make up a big number
and test it to see if it is prime.

How to be Brilliant at Using a Calculator

# Prime magic

In a magic square all rows, columns and diagonals add up to the same number. The magic square below has only prime numbers in it.

Prime numbers are those which can only be divided by 1 and themselves and give a whole number answer.

Complete the magic square. What is the number all the rows, columns and diagonals add up to? Is it a prime number?

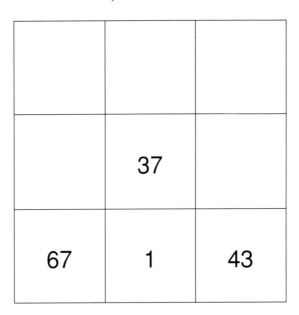

|  |  |  |
|---|---|---|
|  | 37 |  |
| 67 | 1 | 43 |

Make up a magic triangle using these numbers:     **3     5     7     11     13     15.**
The numbers on each side of the triangle should add up to the same number.

Which of these numbers is _not_ prime?

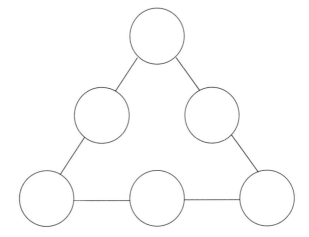

---

**EXTRA!**
Can you make up any other prime magic squares or triangles?

---

# Times table magic

You can amaze your friends by knowing the answer to really difficult times table sums in an instant!

Set up a constant multiplication function to multiply by 5 by pressing either:

| 5 | x | = |

    or

| 5 | x | x | = |

Now enter each number in the table below in turn and press | = |. Record the display each time. *Do not clear the calculator between each go.*

| 2 | 7 | 4 | 11 | 10 | 9 | 3 | 8 | 5 | 12 | 6 | 13 | 20 |
|---|---|---|----|----|---|---|---|---|----|---|----|----|
|   |   |   |    |    |   |   |   |   |    |   |    |    |

What do you think is happening here?

Use this magic method to work out these sums. Make up one of your own in the last column.

|     | 8 x | 13 x | 29 x | ___ x |
|-----|-----|------|------|-------|
| 1   |     |      |      |       |
| 2   |     |      |      |       |
| 3   |     |      |      |       |
| 4   |     |      |      |       |
| 5   |     |      |      |       |
| 6   |     |      |      |       |
| 7   |     |      |      |       |
| 8   |     |      |      |       |
| 9   |     |      |      |       |
| 10  |     |      |      |       |

# Constant division

Set up a constant division function on your calculator. For example, if your number is 2 this can be done either by:

| 2 | ÷ | = |

or

| 2 | ÷ | ÷ | = |

The display will change each time you press | = | .

Record how many times you have to press the | = | key to reach zero for each constant division function.

| Constant function | Number of | = | presses |
|---|---|
| 2 ÷ | |
| 3 ÷ | |
| 4 ÷ | |
| 5 ÷ | |
| 6 ÷ | |
| 7 ÷ | |
| 8 ÷ | |
| 9 ÷ | |
| 1 0 ÷ | |

**EXTRA!**
Can you predict how many key presses it will take to reach zero
with a constant division function for 11? Try it and see.

# Family tally

Work with a friend. You will each need a calculator. One of you should find out how many brothers each person in your class has got. The other should find out how many sisters each person in your class has got. Keep a tally by setting up a constant addition function to add 1 on your calculator. Press $\boxed{=}$ once for each brother/sister.

> **Tip**: To set up an addition function to add 1 press either:
>
> $\boxed{1}$ $\boxed{+}$ $\boxed{=}$        or
>
> $\boxed{1}$ $\boxed{+}$ $\boxed{+}$ $\boxed{=}$

Draw a graph to show your information.

Think of other information you could collect in this way that would fit in with the work you are doing in science, history or geography.

---

### EXTRA!
Compare your results with another class.

---

# Percentages, 1

You can use your calculator to work out percentages. Find the %
key on your calculator. Try the following sequence of keys:

| 2 | 0 | 0 | x | 5 | 0 | % |

What is the answer? What do you think has happened? That is
how you work out what is 50% of 200p.

25% of 200p is
50p. Did you
get it right?

Which keys would you use to work out 25% of 200p?
Try it and see if you were right.

Complete this table:

| Amount | 1% | 10% | 25% | 50% | 75% | 100% |
|--------|-----|------|------|------|------|------|
| 200p | 2p | | | | | |
| 300p | | | 75p | | | |
| 400p | | | | 200p | | |
| 150p | | | | | 112.5p | |
| 250p | | 25p | | | | |
| 350p | | | | | | 350p |

Work with a friend and talk about the things you have noticed about the patterns in the table.

---

**EXTRA!**
Make your own table using some different percentages and different amounts.

---

# Percentages, 2

Practise using the %  key on your calculator. Try the following sequence of keys:

1  0  0  ÷  2  0  0  %

What is the answer? What do you think has happened? That is how you work out what percentage 100p is of 200p.

Decide which keys you would use to work out what percentage 50p is of 200p.
Try it and see if you were right.

Complete this table:

| I had | I spent | I have got now | I spent this % |
|-------|---------|----------------|----------------|
| 200p | 10p | 190p | |
| 200p | 30p | | |
| 200p | 100p | | 50% |
| 200p | 150p | | |
| 300p | 45p | | |
| 300p | 150p | | |
| 300p | 240p | 60p | |
| 300p | 270p | | |
| 400p | 4p | | |
| 400p | 40p | | 10% |
| 400p | 25p | | |
| 400p | 200p | | |
| 400p | 400p | | |

**EXTRA!**
Make up some more calculations of your own. Write some questions to
solve about your class. For example, what percentage of children have
their birthday in June or what percentage of children like chocolate ice-cream?
You may need to round the answer to the nearest percent.

# The restaurant

The restaurant makes a 10% service charge on the total of the bill. Total the following bill and add the service charge.

## Your bill from *Our Restaurant*

| | | |
|---|---|---|
| 2 soups | £1.45 each | £2.90 |
| 1 grapefruit | £0.99 each | £0.99 |
| 3 egg and chips | £2.50 each | £7.50 |
| 2 ice-creams | £1.10 each | £2.20 |
| 1 fruit salad | £1.26 each | £1.26 |
| 3 colas | £0.45 each | £1.35 |

Subtotal _____

plus 10% service charge

TOTAL _____

Make up your own restaurant bill and add a 15% service charge using the quick method.

**Tip**: A quick way to add on 10% is to enter the amount in your calculator then press these keys:

## EXTRA!
VAT is $17\frac{1}{2}\%$ at the moment. To add on VAT enter the amount in your caculator and press these keys:

Look at a price list where VAT hasn't been added (such as in a computer magazine). Use your calculator to work out the actual selling price.

# The sale

You can use your calculator to take a certain percentage off a price. For example, to take off 10%, enter the price then press these keys:

| − | 1 | 0 | % |

Work out the sale prices for these items:

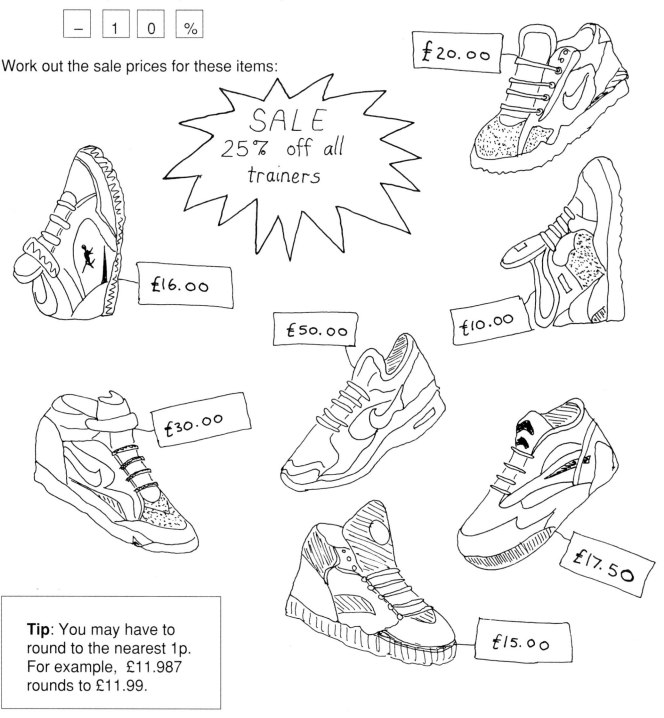

SALE
25% off all
trainers

£20.00

£16.00

£50.00

£10.00

£30.00

£17.50

£15.00

**Tip**: You may have to round to the nearest 1p. For example, £11.987 rounds to £11.99.

## EXTRA!
Use a mail order catalogue to create a small leaflet of goods and prices.
Reduce the prices by 15%.

# Marbles

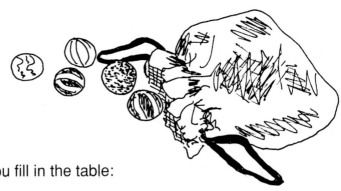

In each packet of marbles:

♦ 30% of the marbles are red
♦ 10% of the marbles are blue
♦ 45% of the marbles are green
♦ 15% of the marbles are yellow

Use the %  key on your calculator to help you fill in the table:

| Total number of marbles | Red marbles | Blue marbles | Green marbles | Yellow marbles |
|---|---|---|---|---|
| 60 | 18 | | | |
| 160 | | | | |
| 360 | | | | 54 |

Check the total number of marbles is correct for each row.

Change the percentage of each colour marble in the packets, then complete this table.
Check the total number of marbles is correct for each row.

♦ _____% of the marbles are red
♦ _____% of the marbles are blue
♦ _____% of the marbles are green
♦ _____% of the marbles are yellow

| Total number of marbles | Red marbles | Blue marbles | Green marbles | Yellow marbles |
|---|---|---|---|---|
| 60 | | | | |
| 160 | | | | |
| 360 | | | | |

---

**EXTRA!**
Compare the percentage of types of fruit sweets in tubes and larger packets.
Do they remain the same?

---

# Percentages, 3

It is often easier to compare information if it is given in percentages. For instance, compare these two classes:

Both classes have 14 girls. Will they have the same percentage of girls?

| Class 1 | | % | Class 2 | | % |
|---|---|---|---|---|---|
| Girls | 14 | | Girls | 14 | |
| Boys | 17 | | Boys | 10 | |
| Total | | | Total | | |

Calculate the percentage of boys and girls in each class like this.

| Number of girls or boys | ÷ | Total number of children in the class | % | You may need to round up or down |
|---|---|---|---|---|

Choose four classes in your school. Look at the class lists of children and work out the percentage of boys and girls in each class.

| Class | % | Class | % | Class | % | Class | % |
|---|---|---|---|---|---|---|---|
| Girls | | Girls | | Girls | | Girls | |
| Boys | | Boys | | Boys | | Boys | |
| Total | | Total | | Total | | Total | |

Now work out the total number of boys and girls in the four classes. What percentage of boys and girls is there in total? Compare each class with this total.

Are there any major differences?

# Pie charts

When drawing pie charts to show information it is often easier to use a calculator to work out the percentage for each category and then calculate the number of degrees for each section of the pie chart. Sometimes you need to round to the nearest degree.

| Die thrown | Tally | Total | % | Degrees |
|---|---|---|---|---|
| 1 | IIII | 4 | 8 | 29 |
| 2 | IHL IIII | 9 | 18 | 65 |
| 3 | IHL II | 7 | 14 | 50 |
| 4 | IHL I | 6 | 12 | 43 |
| 5 | IHL IHL IIII | 14 | 28 | 101 |
| 6 | IHL IHL | 10 | 20 | 72 |
| Total | | 50 | 100 | 360 |

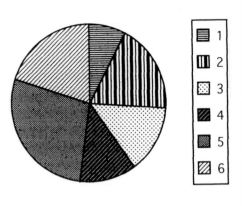

This is the method that was used.

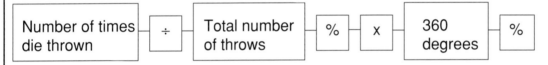

Number of times die thrown ÷ Total number of throws % x 360 degrees %

Try the above calculation for yourself.

Work with a friend and throw a die 50 times for yourself and work out the percentages and number of degrees.

| Die thrown | Tally | Total | % | Degrees |
|---|---|---|---|---|
| 1 | | | | |
| 2 | | | | |
| 3 | | | | |
| 4 | | | | |
| 5 | | | | |
| 6 | | | | |
| Total | | | | |

**EXTRA!**
Work out how many hours a day you spend eating, sleeping, playing and being at school. Make a pie chart using a calculator to help you.

# Square roots

The square root key looks like this $\boxed{\sqrt{\phantom{x}}}$. Enter each number between 1 and 25 in turn and press the square root key. Which numbers give a whole number answer?

| 1 | 2 | 3 | 4 | 5 |
|---|---|---|---|---|
| 6 | 7 | 8 | 9 | 10 |
| 11 | 12 | 13 | 14 | 15 |
| 16 | 17 | 18 | 19 | 20 |
| 21 | 22 | 23 | 24 | 25 |

What do you notice?
What do you think is happening?

Try this sequence:

$\boxed{5}$ $\boxed{\sqrt{\phantom{x}}}$ $\boxed{x}$ $\boxed{=}$

Try some more:

| Starting number | Key presses | | | | Display |
|---|---|---|---|---|---|
| 1 | $\boxed{1}$ $\boxed{\sqrt{}}$ $\boxed{x}$ $\boxed{=}$ | | | | |
| 2 | $\boxed{2}$ $\boxed{\sqrt{}}$ $\boxed{x}$ $\boxed{=}$ | | | | |
| 3 | $\boxed{3}$ $\boxed{\sqrt{}}$ $\boxed{x}$ $\boxed{=}$ | | | | |
| 4 | $\boxed{4}$ $\boxed{\sqrt{}}$ $\boxed{x}$ $\boxed{=}$ | | | | |
| 5 | $\boxed{5}$ $\boxed{\sqrt{}}$ $\boxed{x}$ $\boxed{=}$ | | | | |
| 6 | $\boxed{6}$ $\boxed{\sqrt{}}$ $\boxed{x}$ $\boxed{=}$ | | | | |
| 7 | $\boxed{7}$ $\boxed{\sqrt{}}$ $\boxed{x}$ $\boxed{=}$ | | | | |
| 8 | $\boxed{8}$ $\boxed{\sqrt{}}$ $\boxed{x}$ $\boxed{=}$ | | | | |
| 9 | $\boxed{9}$ $\boxed{\sqrt{}}$ $\boxed{x}$ $\boxed{=}$ | | | | |
| 10 | $\boxed{1}$ $\boxed{0}$ $\boxed{\sqrt{}}$ $\boxed{x}$ $\boxed{=}$ | | | | |

---

### EXTRA!
What are the next five square numbers after 25?
You could record the numbers on a 100 square.

---

How to be Brilliant at Using a Calculator

# Money, money

When you enter money in the calculator you will find that if there is a zero at the end of an amount in £s, for example £2.10, the zero will disappear when you press the next key. Try this:

£2.10      +      £3.40      =      | 5.5 |

The calculator shows 5.5 when it means £5.50.

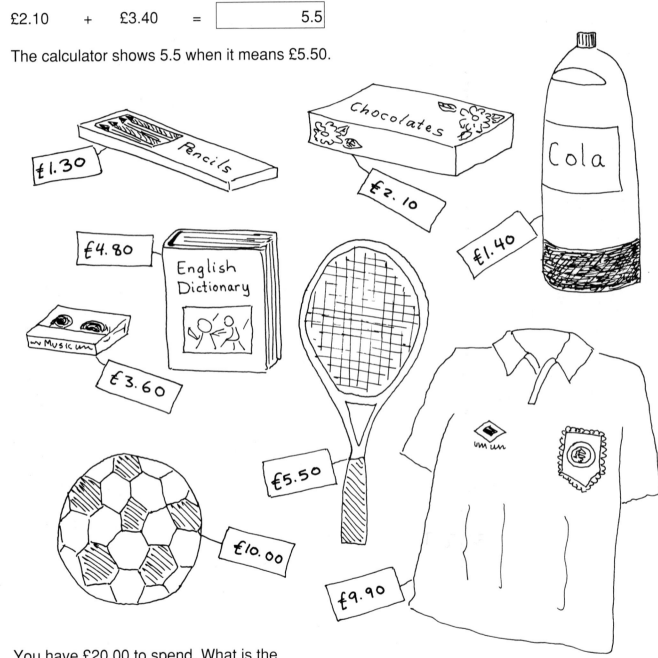

You have £20.00 to spend. What is the largest number of items you could buy?

How much do the two most expensive items cost?

How much do the two cheapest items cost?

**EXTRA!**
Choose two things you would like to buy.
How much do they cost?

# Pennies off

TODAY ONLY!
All prices 10p off!

Use your calculator to work out the new prices.

**Tip**: Remember to use pounds and pence for both amounts or to use pence only for both amounts.

For example:

Frisbee £2.50 – £0.10 = £2.40

or

Frisbee 250p – 10p = 240p

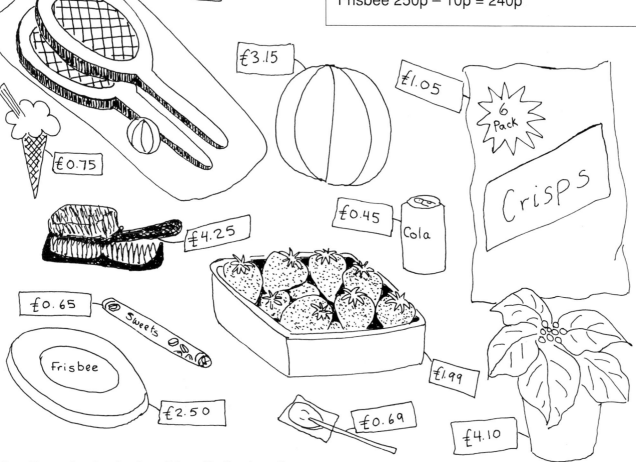

£1.20

£3.15

£1.05

6 Pack

Crisps

£0.75

£0.45   Cola

£4.25

£0.65

Sweets

Frisbee

£1.99

£2.50

£0.69

£4.10

What if you had a further 50p off all prices?

Use your calculator to find out how much it would now cost to buy one of each item.

---

**EXTRA!**
Special today only! An extra 8p off all food and drink!!!

Work out the new prices for the sweets, lollipop, crisps, cola,
ice-cream and strawberries.

---

How to be Brilliant at Using a Calculator

# Memory magic

Calculators usually have a memory. The memory can be used to hold a number you will need later. Follow these flow charts to investigate how the memory works.

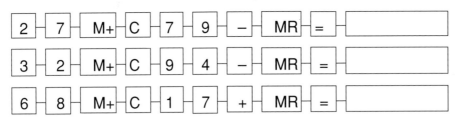

| 2 | 7 | M+ | C | 7 | 9 | − | MR | = | ____ |

| 3 | 2 | M+ | C | 9 | 4 | − | MR | = | ____ |

| 6 | 8 | M+ | C | 1 | 7 | + | MR | = | ____ |

Clear the memory between each sum!

Try a few numbers of your own and see what happens.

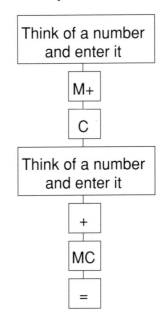

Think of a number and enter it

M+

C

Think of a number and enter it

+

MC

=

Investigate what the | M- | key does. Draw a flow chart showing the steps you take.

**Tip**: The keys for the memory are sometimes labelled differently. Be sure you know what each one does on your calculator.

Recall could be | MR | or | RM | or | $M_C^R$ | or something else.

Clear could be | CM | or | AC | or | $M_C^R$ | or something else.

## EXTRA!
Think about how the calculator's memory could be useful to you.
What type of problems would it help you to solve?

# Going shopping

## Shopping list

5 pints of milk @ 38p
2 packets of butter @ 72p
1 bag of sugar @ 65p
3 loaves of bread @ 68p
2 boxes of eggs @ £1.06
1 jar of jam @ £1.20
4 tins of baked beans @ 35p

You have £11 to spend and want to buy the items on the list. You can work out your change by using the memory on your calculator.

> Enter the money you have to spend into the memory by pressing $\boxed{M+}$.

> Work out the cost of each item by multiplying the number of items by the price of each item you wish to buy.

> Subtract the answer from the memory by pressing $\boxed{M-}$.

> Continue until you have included each item.

> Recall the memory to show what your change would be $\boxed{MR}$.

**Tip**: Remember to use both pounds and pence or pence only.

**EXTRA!**
Make up your own shopping bill and work out your change from £20.

How to be Brilliant at Using a Calculator

# Changing the order

Don't forget to use your calculator!

When subtracting we normally take the smallest number from the larger number, for example:

9    –    2    =    7

What happens if we try it the other way round?

2    –    9    =    ?

Try these:

| | | | | | | | | |
|---|---|---|---|---|---|---|---|---|
| 10 | – | 1 | = | ☐ | 1 | – | 10 | = | ☐ |
| 10 | – | 2 | = | ☐ | 2 | – | 10 | = | ☐ |
| 10 | – | 3 | = | ☐ | 3 | – | 10 | = | ☐ |
| 10 | – | 4 | = | ☐ | 4 | – | 10 | = | ☐ |
| 10 | – | 5 | = | ☐ | 5 | – | 10 | = | ☐ |
| 10 | – | 6 | = | ☐ | 6 | – | 10 | = | ☐ |
| 10 | – | 7 | = | ☐ | 7 | – | 10 | = | ☐ |
| 10 | – | 8 | = | ☐ | 8 | – | 10 | = | ☐ |
| 10 | – | 9 | = | ☐ | 9 | – | 10 | = | ☐ |
| 10 | – | 10 | = | ☐ | 10 | – | 10 | = | ☐ |

Continue the pattern:

| | | | | | | | | |
|---|---|---|---|---|---|---|---|---|
| 10 | – | 11 | = | ☐ | 11 | – | 10 | = | ☐ |
| 10 | – | 12 | = | ☐ | 12 | – | 10 | = | ☐ |
| 10 | – | 13 | = | ☐ | 13 | – | 10 | = | ☐ |
| 10 | – | 14 | = | ☐ | 14 | – | 10 | = | ☐ |
| 10 | – | 15 | = | ☐ | 15 | – | 10 | = | ☐ |

## EXTRA!
### What would the answer to these be?

10    –    20    =    ☐        20    –    10    =    ☐

Try with some different numbers.

How to be Brilliant at Using a Calculator

Name_____

# I can use a calculator to...

Date

| | |
|---|---|
| add numbers together   $+$   $=$ | |
| subtract one number from another   $-$   $=$ | |
| multiply two numbers together   $\times$   $=$ | |
| divide one number by another   $\div$   $=$ | |
| clear the display   $C$ | |
| clear an error   $CE$ | |
| set up a constant addition function   $+$   $=$   $=$ | |
| set up a constant subtraction function   $-$   $=$   $=$ | |
| set up a constant multiplication function   $\times$   $=$   $=$ | |
| set up a constant division function   $\div$   $=$   $=$ | |
| calculate a percentage of an amount   $\times$   $\%$ | |
| calculate a fraction as a percentage   $\div$   $\%$ | |
| calculate a fraction as a decimal   $\div$   $=$ | |
| calculate a square root   $\sqrt{\ }$ | |

# I can ...

| | Date |
|---|---|
| read the display when it shows money | |
| calculate with amounts in pounds and in pence | |
| round to the nearest 1 | |
| round to the nearest 10 | |
| round to the nearest 100 | |
| round to the nearest 1000 | |
| interpret the display sensibly and round up or down as appropriate | |
| recognize if the answer is about the right size | |
| add numbers into the memory | |
| subtract numbers from the memory | |
| recall the memory | |
| clear the memory | |

# 100 square

| 1 | 2 | 3 | 4 | 5 | 6 | 7 | 8 | 9 | 10 |
|---|---|---|---|---|---|---|---|---|---|
| 11 | 12 | 13 | 14 | 15 | 16 | 17 | 18 | 19 | 20 |
| 21 | 22 | 23 | 24 | 25 | 26 | 27 | 28 | 29 | 30 |
| 31 | 32 | 33 | 34 | 35 | 36 | 37 | 38 | 39 | 40 |
| 41 | 42 | 43 | 44 | 45 | 46 | 47 | 48 | 49 | 50 |
| 51 | 52 | 53 | 54 | 55 | 56 | 57 | 58 | 59 | 60 |
| 61 | 62 | 63 | 64 | 65 | 66 | 67 | 68 | 69 | 70 |
| 71 | 72 | 73 | 74 | 75 | 76 | 77 | 78 | 79 | 80 |
| 81 | 82 | 83 | 84 | 85 | 86 | 87 | 88 | 89 | 90 |
| 91 | 92 | 93 | 94 | 95 | 96 | 97 | 98 | 99 | 100 |

# Sequence cards

| Enter a<br>number | Record<br>the result | Multiply by 4 |
|---|---|---|
| Subtract 5 | Divide by 3 | Add 0 |
| Multiply by 1 | Divide by 1 | Add 1 |
| Subtract 1 | Multiply by 0 | Divide by 10 |
|  |  |  |
|  |  |  |

How to be Brilliant at Using a Calculator